Asa Gray

Botany for young people

How plants behave: how they move, climb, employ insects to work for them

Asa Gray

Botany for young people
How plants behave: how they move, climb, employ insects to work for them

ISBN/EAN: 9783741170737

Manufactured in Europe, USA, Canada, Australia, Japa

Cover: Foto ©Klaus-Uwe Gerhardt /pixelio.de

Manufactured and distributed by brebook publishing software
(www.brebook.com)

Asa Gray

Botany for young people

BOTANY For YOUNG PEOPLE

HOW PLANTS BEHAVE

𝔅𝔬𝔱𝔞𝔫𝔶 𝔣𝔬𝔯 𝔜𝔬𝔲𝔫𝔤 𝔓𝔢𝔬𝔭𝔩𝔢.

PART II.

HOW PLANTS BEHAVE:

HOW THEY MOVE, CLIMB, EMPLOY INSECTS TO WORK FOR THEM, &c.

By ASA GRAY.

NEW YORK ·:· CINCINNATI ·:· CHICAGO

AMERICAN BOOK COMPANY

PREFACE.

HOW PLANTS GROW, the first part of BOTANY FOR YOUNG PEOPLE AND COMMON SCHOOLS, was written fourteen years ago, in the endeavor to provide a book upon Elementary Botany, adapted to the instruction of young people, even of children, yet truly presenting, albeit in a simple way, the leading facts, methods, and principles of the science as understood by its masters. The book has been successful. It will probably enable a young person, under the guidance of a qualified teacher, to obtain a larger, truer, and worthier knowledge of Botany than many grown people could readily find the way to acquire a generation ago.

That young people, that all students, indeed, should be taught to observe, and should study Nature at sight, is a trite remark of the day. But it is only when they are using the mind's eye as well, and raising their conceptions to the relations and adaptations of things, that they are either learning science or receiving the full educational benefit of such a study as Botany or any other department of Natural History.

There is a study of plants and flowers admirably adapted, while exciting a lively curiosity, to stimulate both observation and thought, to which I have long wished to introduce pupils of an early age. The time has now arrived in which I may make the attempt, and may ask young people to consider not only 'How Plants Grow,' but How Plants Act, in certain important respects, easy to be observed, — everywhere open to observation, but (like other common things and common doings) very seldom seen or attended to. This little treatise, designed to open the way for the young student into this new, and, I trust, attractive field, may be regarded as a supplement to the now well-known book, the title of which is cited at the beginning of this prefatory note. If my expectations are fulfilled, it will add some very interesting chapters to the popular history of Plant-life.

Although written with a view to elementary instruction, and therefore with all practicable plainness, the subjects here presented are likely to be as novel, and perhaps as interesting, to older as to young readers.

To those who may wish to pursue such studies further, and to those who will notice how much is cut short or omitted (as, for instance, all reference to discoverers and to sources of information), I may state that I expect to treat this subject in a different way, and probably with somewhat of scientific and historical fulness, in a new edition of a work intended for advanced students.

A.

BOTANIC GARDEN, HARVARD UNIVERSITY,
February 20, 1872.

Vignette Title-Page. — Left-hand side, an Ivy climbs by rootlets and a Passion-flower by tendrils ; right-hand, a Nepenthes by pitcher-bearing tendrils, and a Morning-Glory by twining stem : bottom, at the left of the centre, a Rhodochiton, and at the right a Maurandia climb by their leafstalks. Bottom, left-hand side, a Green Orchis (Habenaria orbiculata) sends up from between a pair of large round leaves a raceme of long-spurred flowers. Two Orchid Air-plants at the top, viz., Stanhopea tigrina at the centre, a Phalænopsis at the right-hand corner. Two leaves of Sarracenia rubra, an American Pitcher-plant, rise from near the lower right-hand corner : in front of them is a Sundew, Drosera rotundifolia ; at the centre a Venus's Fly-trap, Dionæa muscipula.

HOW PLANTS BEHAVE.

CHAPTER I.

HOW PLANTS MOVE, CLIMB, AND TAKE POSITIONS.

1. Two plants — one of them common in cultivation, and the other rarer, but almost as easy to raise — are looked upon as vegetable wonders, namely, the Sensitive Plant and Desmodium gyrans. They are striking examples of

2. **Plants that move their Leaves freely and rapidly.** In the well-known Sensitive Plant (*Mimosa pudica*) the foliage quickly changes its position when touched, appearing to shrink away from the hand. Fig. 1 represents a piece of stem with two (compound) leaves; the lower one expanded, as it is in sunshine and when untouched : the upper leaf shows the position which is taken, by quick movements, when roughly brushed by the hand. It makes three movements. First, the numerous leaflets close in pairs, bringing their upper faces together and also inclining forwards; then the four branches of the leafstalk, which were outspread like the rays of a fan, approach each other; at the same time the main leafstalk turns downward, bending at its joint with the stem. So the leaf (for it is all one compound leaf) closes and seemingly collapses at the touch. In a short time, if left to itself, it slowly recovers the former outspreading position.

3. The second plant, *Desmodium*

Fig. 1. Sensitive Plant.

gyrans (we have no common name for it), also belongs to the great Pulse Family, and flourishes in warm climates. It inhabits the warmer parts of India, but is easy to cultivate in a hot-house, or even in an open garden during the heat of summer. The leaves are of only three leaflets (Fig. 2), a large one at the end of the leafstalk, accompanied by a pair of small leaflets, one on each side. The

Fig. 2. Desmodium gyrans.

end leaflet usually moves too slowly to be seen, and only as light is given or withdrawn; we have seen it move rather briskly, however, upon one occasion. The side leaflets are active enough. Under the temperature of a sultry summer's day they may be seen to rise and fall by a succession of jerking movements, not unlike those of the second-hand of a clock, but without much regularity, now stopping for some time, then moving briskly, always resting for a while in some part of their course, commonly at the highest and lowest points, and starting again without apparent cause, seemingly of their own will. The movement is not simply up and down, but the end of the moving leaflet sweeps more or less of a circuit. It is not set in motion by a touch, but begins, goes on, or stops of itself.

4. Whether these movements are of any use to these plants is more than we can tell; nor do we very well know how they are effected. The attempts that have been made to explain how the motion is brought about need not be considered here. However done, it is clear that the *leaves move by their own act*, — in the one case responding to a touch; in the other independently, or, as we say, spontaneously.

5. Now, truly wonderful as these two plants are, there is nothing really peculiar about them. By which is meant, not merely that some other plants are known to move as freely, though perhaps less rapidly, but that many ordinary plants perform similar movements, in one or both of these ways, and that all plants possess similar faculties. The hour-hand of the clock moves as really as the minute-hand and the second-hand, although the motion of the latter only is

discerned by the eye. Lifeless things may be moved or acted on; living beings move and act, — plants less conspicuously, but no less really, than animals. In sharing the mysterious gift of life, they share some of its simpler powers.

6. **The Sleep of Plants,** as Linnæus fancifully termed it, — that is, the different position which leaves and leaflets take at nightfall, — is a familiar case of free movement, only the motion is too slow to be seen by the eye. The Sensitive Plant is a good instance of this. Its leaves slowly assume the same posture at or before sunset that they rapidly do when disturbed by a touch or jar, and they remain so until the light of morning. Most other plants of the Pulse Family (the Locusts, for instance), and many of other families, take a very different position by night from that of day. The end-leaflet of Desmodium gyrans hangs down as soon as the light of day begins to wane, but rises and turns its upper face to the sun again in the morning.

7. **The Turning of Green Shoots to the Light,** which we observe when house-plants are kept in our windows, and the turning of the upper face of most leaves towards the lighted side, are similar cases of slow movement or bending. Many people suppose that the green shoot *grows* towards the light, whereas it only *bends* towards it. One has only to notice the behavior of the slender stemlet of a seedling Radish, or of any similar plant, when set in a window, and see it bending towards the lighted side in a few minutes, before it has had time to grow perceptibly, to be convinced that the growth and the bending are different acts.

8. **The contrary Directions of Stem and Root** when springing from the seed are of this kind. Read the brief account given in 'How Plants Grow,' paragraphs 28 and 29, and watch the operation in young seedlings. Note how one end of the embryo plantlet rises out of the soil and into the light, and, if need be, turns quite round to do so, while the other turns from the light and strikes deeper into the ground. This shows that it is the plant itself which acts in taking these directions, and that these positions are the result of real movements, however slow.

9. **Climbing Plants** afford some of the most curious and most varied illustrations of the movements which plants perform; and in these it is easy to see what the movements are for. The advantage which a plant gains by climbing is, that it may thereby rise higher and get a fuller exposure to the light than it could with the same amount of material if it stood independently. Compare the amount of wood or other material in a tree with that of any climber which has ascended it and made a support of its topmost branches. Plants climb in several ways. Some are

10. **Root-Climbers.** These creep up the face of rocks or walls, or the trunks of trees, their stems, as they grow, pressing against the support and adhering to it by means of numerous rootlets which they throw out: the end of these rootlets commonly flattens out or expands into a small disk or holdfast which adheres to the wall or bark, etc. Ivy, that is, true or "English" Ivy, is a good example of this. See the vignette title-page, left-hand side. Our Poison Ivy and the Trumpet Creeper climb in the same way. There is, perhaps, no more effectual mode of climbing when bare walls or large trunks are the support. In other cases

11. **Twiners, i. e. Twining Plants,** have an obvious advantage. To twine spirally round some supporting body is a common mode of climbing. This is done by a movement of the stem itself, not less remarkable in reality than that of the leaflets of the Desmodium gyrans, just described, and indeed of similar nature. The Hop and some Honeysuckles twine with the sun. Morning Glory, and all the Bindweeds of the Convolvulus Family, Beans, and indeed most of the common twiners, turn against the sun, that is, from the left to the right hand of the observer.

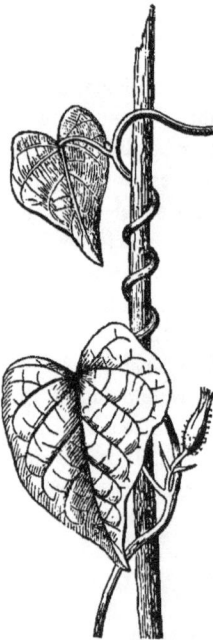

12. When a twining stem overtops its support, the lengthening shoot is seen thrown over to one side, and usually outstretched, as in Fig. 3. One might suppose it had fallen over by its weight; but it is not generally so. If turned over, say to the north, when first observed, it will probably be found reclining to the south an hour or so later, and an hour later again turned northward. That is, the end of the stem is sweeping round in a circle continually, like the hand of a clock. It keeps on growing as it

Fig. 3. Morning Glory, twining. revolves; but the revolving has nothing to do with the growth, and, indeed, is often so rapid that several complete sweeps may be made before any increase in length could be observed. The time of revolving varies in different species. It also depends upon the weather, being slow or imperceptible when it is cool, and more rapid when it is warmer. Sometimes it stops when everything seems favorable, and starts again after a while. The Hop, Bean, and Morning Glory are as quick as any. In a sultry day, and when in full vigor, they commonly sweep round the circle in less than two hours. They move by night as well as by day. When the free summit of a twining stem is outstretched to two feet or more in length, so as to magnify the motion, this is sometimes rapid enough to be actually seen in some part of the circuit.

13. Because twining stems are often twisted more or less, some have supposed that the twisting was the cause of the revolving sweep of the free end. If so, the stem below would in a day or two be likely to twist itself off. And twiners seldom twist much when climbing a smooth and even support. To learn how the sweeps are made, one has only to mark a line of dots along the upper side of the outstretched revolving end of such a stem (say that of the Morning Glory, Fig. 3), and to note that when it has moved round a quarter of a circle, these dots will be on one side; when half round, the dots occupy the lower side; and when the revolution is completed, they are again on the upper side. That is, the stem revolves by bowing itself over to one side, — is either pulled over or pushed over, or both, by some internal force, which acts in turn all round the stem in the direction in which it sweeps; and so the stem makes its circuits without twisting.

14. So the sweeping round of the stem is a movement like that wonderful one of the leaflets of Desmodium gyrans, just described, only slower. And here we see what it is for. The sweeping movement of the stem is the cause of the twining. The stem sweeps round that it may reach some neighboring support; as it grows it sweeps a wider and wider space, that is, reaches farther and farther out. When it strikes against any solid body, like the stalk of a neighboring plant, it is stopped : but the portion beyond the contact is free to move as before; and, continuing to lengthen and to move on, it necessarily winds itself round the support, that is, *twines*. This is the explanation of twining climbers.

15. **Leaf-Climbers.** Some plants climb by their leaves, either the blade, or more commonly the petiole, hooking or coiling round something within reach. *Clematis* or Virgin's-Bower is a familiar instance. In all the common species of Clematis the leaves are compound, and the divisions of the petiole, or at first the young leaflets themselves, bend round the stalks or branches of neighboring plants, or any supporting object not too large to be grasped, and so ascend. *Lophospermum* and *Maurandia* (handsome flowering herbs of the gardens), and one or two other

plants of the same family, with simple leaves, climb freely in this way, neatly coiling their leafstalk round any slender support within reach.   The vignette title-page shows two illustrations of this, in the lower part.

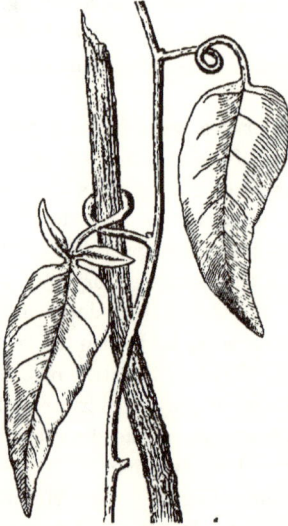

Fig. 4.   Solanum jasminoides, climbing by its leafstalks.

16. A rather common cultivated species of Nightshade, *Solanum jasminoides*, is a good example of the same kind, and furnishes the present illustration, in Fig. 4.   It is interesting to notice how the leafstalks of this plant which have clasped a support grow much stouter and firmer than those which have not, becoming three or four times as thick as before, — as if the need of greater strength and rigidity somehow brought it about.

17. A leaf-climber has this advantage over a twiner, that it may reach a given height with less amount of substance.   Its stem may rise straight up, and save much in length over the twiner, which has to produce twice or thrice that length of stem in reaching the same elevation, on account of the coils.

18. To understand how leaves or leafstalks lay hold of a support, we must refer back to the Sensitive Plant (Paragraph 2); its leaves and leafstalks, we know, respond to the touch of a foreign body by a movement.   So do those of leaf-climbers : only the movement by which they clasp the support is very slow and incited only by prolonged contact.   If one of these leafstalks be rubbed for some time with a piece of wood, it will generally respond to the irritation by curving; but it will be two or three days about it; and in two or three days more it may straighten itself, unless the stick is left in contact with the leafstalk : then it will clasp it permanently, making one or perhaps two turns around it, and in time it may thicken and harden.   That the climbing in such cases is the result of a movement, however slow, under sensitiveness to touch, is further shown by the behavior of tendrils.

19. Between leaf-climbing and tendril-climbing there is every gradation.   In *Gloriosa*, a tropical plant of the Lily Family, the tip of a simple leaf extends

into a slender hook, for laying hold of anything within reach. In *Nepenthes* (a climbing sort of Pitcher-plant, shown on the right-hand side of the vignette title, and one leaf in Fig. 5, on a larger scale), the tip of the blade grows out into a tendril which acts just as does the leafstalk of Fig. 4 and of the other leaf-climbers; at the end of this a pitcher, with a lid to it, is generally formed. Of this more is to be said hereafter. In that vigorous climber, *Cobœa*, the branching claws and grapples which are used to such effect are merely the upper portion of a compound leaf changing into tendrils. The tendrils of a Pea are similar, but simpler.

20. **Tendril-Climbers** are best illustrated by such plants as Passion-flowers (see vignette title, on the left, and Fig. 6): here the tendril is a simple thread-like shoot, for the purpose of climbing and nothing else. This is the most exquisite, and under many circumstances the most advantageous, as it is one of the commonest of the contrivances for climbing. The tendril, as it grows, stretches out

Fig. 5. Leaf of Nepenthes.

horizontally, as if in search of a supporting object. More slender than a stem or any other sort of stalk, it can thus extend farther at the least expense of material.

21. In the most perfect tendrils, and notably in the slender Passion-flowers (such as the annual *Passiflora gracilis*, and the Maple-leaved species, *P. acerifolia*, Fig. 6), opportunities for securing a hold are much increased by the revolving of the tendril. It sweeps circuits, like the stem of a twiner, although with less regularity, sometimes, however, with greater rapidity. In hot weather these tendrils often move through the complete circle in an hour or less, or even so fast that the motion of the end of a long tendril may sometimes be distinctly seen in a part of its course. The revolving of tendrils is more fitful than that of twining stems: they often stop for a while, or move very slowly or irregularly. Some tendrils, as we shall soon see, do not revolve at all.

22. If a tendril does not reach anything, after attaining its full growth and remaining for some time outstretched, it then either coils up from the end (as seen in the middle tendril of Fig. 6), or else becomes flabby, hangs down in an exhausted state, dies, and withers away.

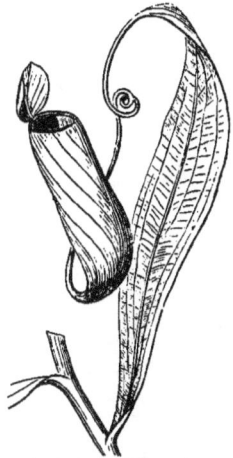

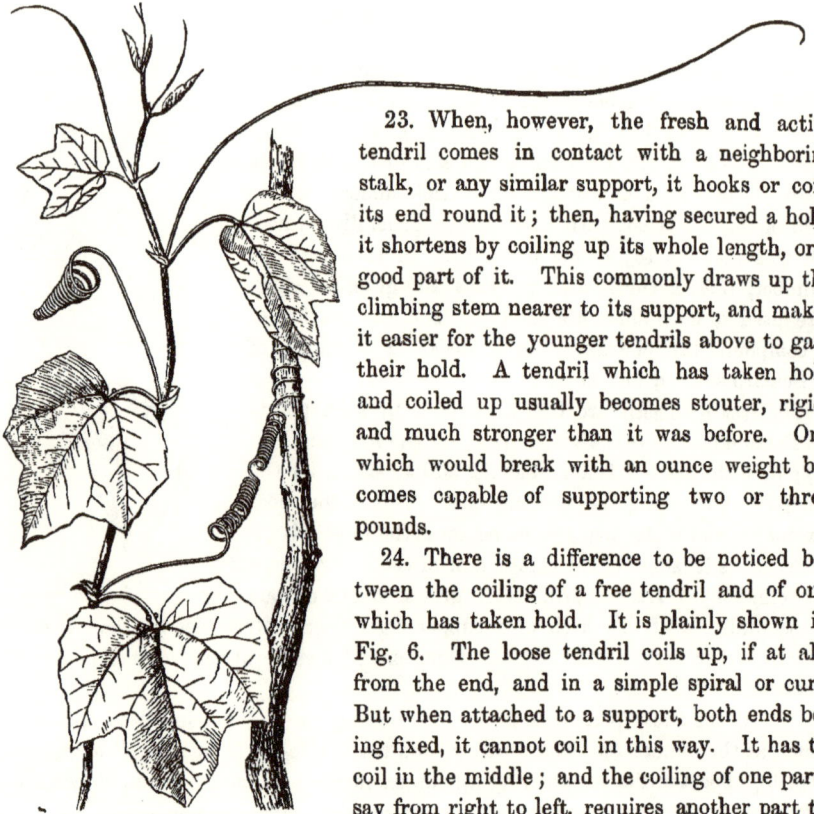

Fig. 6. Maple-leaved Passion-flower, with ten-drils in various states.

23. When, however, the fresh and active tendril comes in contact with a neighboring stalk, or any similar support, it hooks or coils its end round it; then, having secured a hold, it shortens by coiling up its whole length, or a good part of it. This commonly draws up the climbing stem nearer to its support, and makes it easier for the younger tendrils above to gain their hold. A tendril which has taken hold and coiled up usually becomes stouter, rigid, and much stronger than it was before. One which would break with an ounce weight becomes capable of supporting two or three pounds.

24. There is a difference to be noticed between the coiling of a free tendril and of one which has taken hold. It is plainly shown in Fig. 6. The loose tendril coils up, if at all, from the end, and in a simple spiral or curl. But when attached to a support, both ends being fixed, it cannot coil in this way. It has to coil in the middle; and the coiling of one part, say from right to left, requires another part to twist as much in the opposite direction. So the coil has a break in the middle, half twisting one way and half the other way, as is shown in the lower tendril of the figure. A longer tendril often has three or four, or even five or six, such breaks, the portions coiled successively in opposite directions.

25. Pumpkins, Squashes, and all the Gourd Family furnish excellent examples of these actions of tendrils. Their tendrils are like those of Passion-flowers, except that they are mostly branched or compound, and, like the claws of a bird, stretch out in several directions.

26. There is great variety in the behavior of different tendrils. Those of the Grapevine do not make sweeps, but stretch out away from the light, or in the direction from which least light comes, — an instinct which is apt to lead them to a support, — and the two forks diverge, as if feeling for something to lay hold of. When they reach anything that can be surrounded, one fork commonly grasps from one side, the other from the opposite side, somewhat as an object would be grasped by a thumb and finger.

27. The more branching tendrils of the Virginia Creeper equally turn from the light, and therefore towards the wall or trunk, which this climber delights to occupy and cover. When their tips reach the wall they expand into a disk or flat plate, which adheres firmly to the surface. This particularly adapts the Virginia Creeper to ascending walls or other flat surfaces. The tendrils which do not attach themselves remain slender, and in a week or two shrink and wither away.

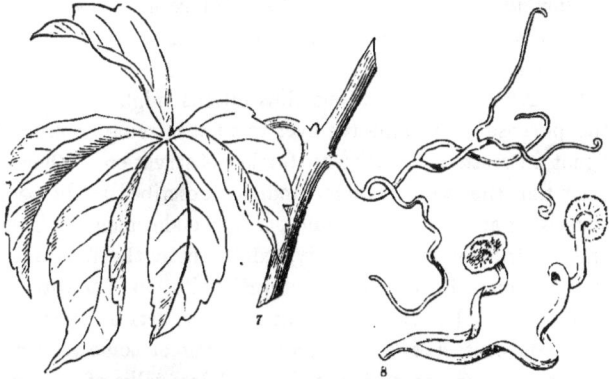

Fig. 7. Virginia Creeper: tendril beginning to form its disks or holdfasts. 8. Older branches with full-formed disks.

Those that do usually spread their branches widely apart, like fingers of an outstretched hand, form their disks and fix them fast to the wall; then they contract more or less into coils, and at length grow stronger and more rigid; so that they last for years, and endure a pretty heavy strain without breaking or parting from the wall. It is most interesting to see how the strain is divided by these five or six separate attachments, by the coiling of each branch to give elasticity, so that the pull shall come upon all at once, and to note the strengthening of the whole by the formation of more woody fibre. The strain is distributed among the branches, and the whole combination is so strong that it is rarely torn away by wind or storm.

28. In revolving tendrils the most wonderful thing to remark is the way in

2

which they avoid winding themselves around the stem they belong to. The active tendrils are of course near the top of the stem or branch. The growing summit beyond the tendril now seeking a support is often turned over to one side, so that the tendril, revolving almost horizontally, has a clear sweep above it. But as the growing stem lengthens and rises, the tendril might strike against it and be wound up around it. It never does. If we watch these slender Passion-flowers, which show the revolving so well in a sultry day, we may see, with wonder, that when a tendril, sweeping horizontally, comes round so that its base nears the parent stem rising above it, it stops short, *rises stiffly upright, moves on in this position until it passes by the stem, then rapidly comes down again to the horizontal position,* and moves on so until it again approaches and again avoids the impending obstacle !

29. Other equally curious illustrations might be given; but these may serve the purpose of opening the eyes to what is going on around us, awaken an intelligent interest, and excite to further observation. They are enough to make it clear that the two vegetable prodigies described at the beginning of this chapter, surprising as they are, have no peculiar endowments. Climbing plants generally, and tendril-climbers especially, exhibit both the free movements of the one, and the movement in response to external irritation of the other. The sweeping round of tendrils is like the movement of the leaflets of Desmodium gyrans : their coiling upon contact, and the similar coiling of some leafstalks, are to be compared with the movement of the leaflets and leafstalks of the Sensitive Plant.

30. This becomes evident when the motion is quick enough to be seen by the eye. It has already been stated that a very long tendril of one of the slender Passion-flowers has often been seen to move. Still oftener may it be seen to coil up at the tip when gently rubbed. This is also to be seen in the Bur-Cucumber (*Sicyos*), a common weed of the Gourd Family. When, in a sultry summer day, we gently rub, with a stick or with the finger, the upper end of a vigorous tendril, it will respond within half a minute by coiling up so rapidly that the motion may be distinctly seen. It will soon straighten, but will coil again if the rubbing is repeated. If a stick be left in contact the coiling will be permanent; and a downward propagation of the same action is what throws the whole tendril into spiral coils.

CHAPTER II.

HOW PLANTS EMPLOY INSECTS TO WORK FOR THEM.

31. PLANTS supply animals with food. That, we may say, is what they were made for. In some cases the whole herbage, in others the fruit, seeds, bulbs, tubers, or roots, are fed upon. But vast numbers of insects, and some birds (such as humming-birds), draw nourishment from plants, mainly from their flowers, without destroying or harming them. By their colors, odors, and nectar, blossoms attract insects in great numbers and variety.

32. Nectar, the sweet liquid which most flowers produce, is the real attraction : bright colors and fragrance are merely advertisements. This sweet liquid is often called honey ; but *nectar* is the proper name for it, as it is not really honey until it is made so by the bee. Some insects also take pollen (the powdery matter produced in the anthers : see How Plants Grow, paragraph 17), either for their own consumption or that of their progeny. That may possibly do the plant some harm. But the nectar they consume is of no use to flowers that we know of, except it be to entice insects.

33. So flowers are evidently useful to insects, and most flowers are feeding-places for them. Where free lunches are provided some advantage is generally expected from the treat : and we are naturally led to inquire,

34. **Why should Flowers entice Insects to visit them?** What advantage are they likely to derive in return for the food they offer ? In certain cases the use of insects to flowers is evident enough. When, in early spring, we see Willow-catkins thronged with honey-bees, and notice that their blossoms are of the separated sort (How Plants Grow, 205), — those of one tree consisting of stamens only, of another tree, of pistils only, — and that the bees flying from tree to tree have their bodies well dusted with pollen, we may conclude that the bees are doing useful work in carrying pollen from the stamen-bearing flowers that produce it to the pistil-bearing flowers that require it in order to set seed (see How Plants Grow, 16, 196). While feeding from the stamen-bearing catkins, their heads and bodies, rubbing against the anthers, get dusted with the pollen. When they fly to a

tree with pistil-bearing catkins, some of this pollen is rubbed upon the stigmas, and in consequence its fruit may set and the seeds be perfected. The stamens and pistils of Willows being on different trees, and the two sorts of trees very likely at a wide distance apart, it is necessary that the pollen should be carried by insects or some other conveyance, if the Willow is to be propagated by seed.

35. It might have been left to the winds to waft the pollen. It is so in Pine-trees, Spruces, and the like. But considering what enormous superabundance of pollen these trees produce (even when the two sorts of flowers are on the same tree) in order to make sure of the result, one cannot doubt that there is great ' economy in the arrangement by which the busy bees are called upon to do the carry-ing. In such instances the insects are probably as useful to the flowers as the flowers are to the insects.

36. **Why should perfect Flowers need to attract Insects?** Far the larger number of flowers are *perfect*, that is, are furnished with both stamens and pistils : the sta-mens are almost always more numerous than the pistils, and encompass them ; and each anther contains a thousand or many thousand times more grains of pol-len than there are of seeds to be fertilized, and all so near or in such position that it appears as if the pollen, or a sufficient quantity of it for the purpose, must needs be shed upon the stigmas, either with or without the aid of the wind. Yet here insects, in searching the blossoms for food, might be helpful even if not needful.

37. There are plenty of flowers, however, to which insects could seemingly be of no use. They have stamens and pistils not only close together, but even in contact, — shut up together in some cases, so that some of the pollen cannot fail to be shed upon the stigma. Pea-blossoms, and those of most of the Pulse Fam-ily are examples of this, having ten anthers closely surrounding one stigma, and enclosed by a pair of the petals. And in the Showy Dicentra (or Bleeding-heart, as it is popularly called, from the shape and color of the corolla), as in all the rest of the Fumitory Family, six anthers are completely enclosed with one stigma, three on one side and three on the other, in a cavity just large enough to hold them. This cavity is formed by the spoon-shaped summits of the two inner petals, which never separate, being united only at their tips : those of the two outer and larger petals open and turn back. (See Figs. 9, 10.) One would say that such blossoms are purposely and effectually arranged to be fertilized without any assistance, and to exclude all interference by insects. Yet they produce nectar

and are visited by bees.   Is their nectar provided only for the good of the bee?
We might suppose so, until we come to know the remarkable fact that, unless
visited by insects, they seldom ripen a pod or set a seed.   The Showy Dicentra,
which comes from Japan or Northern China, rarely sets fruit in our gardens in
any case.   But the wild species of Corydalis and Fumitory, which have their
flowers on the same plan, seed freely enough.   Yet when the blossoms are kept
covered with fine gauze, so as to exclude insects, little or no seed is produced.
Evidently then, for some reason or other, insects sucking their honey are not only
useful, but needful even to such blossoms.   Why they should be needful remains
to be seen.

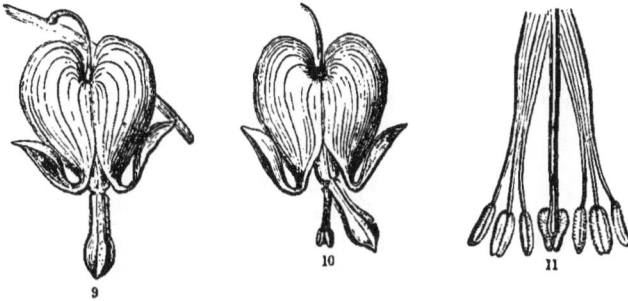

Fig. 9. Flower of Bleeding-heart, Dicentra spectabilis   Fig. 10 Same, with the tips of the united inner petals pushed
to one side.   Fig. 11. Tips of the six stamens and pistil, which are exposed in Fig. 10, here separated and dis-
played, magnified.

38. If it be wonderful that such flowers as the last do not well fertilize without
help, although constructed, as we should say, expressly to do it, equally wonderful
is it to find blossoms with anthers and stigma placed close together, but with some
obstacle interposed, as shown on near examination; which looks as if the object
were *how not to do it.*

39. Iris-flowers are of this sort.   There is a stamen to each of the three stigmas, and close beside it.   Behind each stamen and partly overhanging it is a
petal-like body, peculiar to Iris or Flower-de-Luce : these three bodies, appearing
like supernumerary petals, are divisions of the style, in a peculiar form, notched
at the end ; under the notch is the stigma, in the form of a thin plate.   We
notice that the stigma is higher than the anther ; but that is only a part of the

difficulty. The anther and the stigma face away from each other. The anther faces outwards and discharges its pollen through two long slits on the outer side only. The thin plate or shelf is stigma only on its upper or inner face, which is roughened and moistened in the usual way for receiving the pollen: the face turned towards the anther cannot receive the pollen at all.

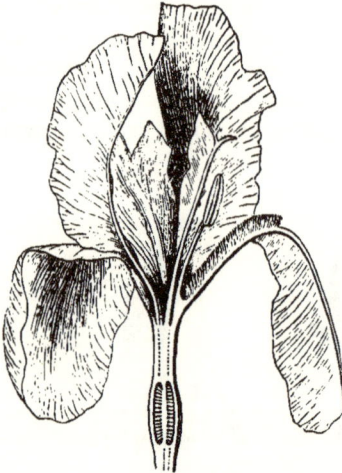

40. A less common flower, the beautiful *Arethusa*, of our northern bogs (Figs. 13, 14), is quite as curiously arranged so as *just not to do* of itself what is obviously intended to be done. The stamen and the style are united into a long and wing-margined column; the stigma is a shelf; and the anther, which is shaped like a helmet, and is fixed to the top of the column by a hinge at the back, rests upon this shelf, its front edge at bottom projecting slightly over its edge, — just as the lid of a chest projects a little over the front side, for more convenient lifting. The anther holds four soft and loose pellets of pollen, which are ready to fall out when the anther is uplifted. But here again, only the under side of the shelf is actually stigma; the pollen lies imprisoned on the upper surface, and can never of itself reach the lower surface, where alone it can act.

Fig. 12.   Iris-flower cut lengthwise, showing one stamen and stigma.

41. There are hundreds of such cases, differing more or less in the arrangement, but agreeing in this, that the pollen is placed tantalizingly near the stigma, yet where it can never reach it of itself, or can seldom and only accidentally do so. Surely, if we had the making of these blossoms, we should have turned the shelf under the anther of Arethusa the other side up, and have restored the harmony of that averted couple in Iris by turning the two face to face in place of back to back.

42. The flower of *Aristolochia Sipho*, or Pipe-vine of the Southern States (a large-leaved woody twiner which is cultivated for arbors), shows the same extraordinary aversion in a different way. From its shape the blossom is called Dutchman's-pipe : it is a tube curved round on itself, largest at the base, contracted at

the orifice, and then expanded into a flat border. At the very bottom of it is a short and thick mass, consisting of a broad stigma, to the outside of which three sets of anthers grow fast : these face away from the stigma, so that none of the pollen can fall on it; and the crooked tube of the flower, with a narrow opening, must effectually prevent the wind from giving any aid. What can this mean ?

43. To explain the puzzle which such flowers present, we have to consider that, by their bright colors, or odors, or the nectar they offer, — sometimes by all three allurements combined, always by the latter, — they attract insects; by whose usually rough or bristly heads, or legs, or bodies, pollen may be brushed out of the anthers, or caught as it falls, and some of it carried to or dropped upon the stigma. And we must infer that these blossoms are so constructed and arranged on purpose that insects may visit and fertilize them; and that many species are absolutely dependent upon such assistance : for, as they would not set seed, they could not permanently exist, except for the insects which they nourish in return for such service. So we conclude that honey is the wages paid to insects in return for the work they do ; and that the fragrance of flowers and their beautiful colors, as well as their honeyed sweets, are not merely for our delight, and for the use of the insects they feed, but are of primary use to the plant itself.

44. In confirmation of this view, it is found that flowers which are fertilized by the wind, of which there are numerous sorts, produce neither bright-colored corollas, nor fragrance, nor honey.

45. Now that we know the way of it, nothing is more interesting than to notice how particular flowers, each in its own way, are arranged so as to be helped by the insects that visit them. Iris-flowers (Fig. 12), for instance, are visited by bees. These alight upon the outer and recurving, usually crested or bearded divisions of the flower, down the base of which is the only access to the nectar below. When sucking out the nectar with its proboscis, the bee's head is brought down beneath the anther; when raised, it will rub against it and brush out some of the pollen : this, loosely adhering to its hairy surface, is ready to be deposited upon the shelf of stigma above, not when the bee leaves the flower, but when it repeats the action. When Arethusa (Figs. 13 – 15) is visited, the head of the bee enters the mouth of the flower : in raising it to leave the flower after extracting the nectar, the head hits the front edge of the helmet-shaped anther, raises it like a lid, and receives one or more of the soft pellets of pollen that fall upon it : on again entering the flower and again rising to depart, the pollen-loaded head is first

brought against the sticky stigma, which occupies all the lower face of the shelf, and at the next instant raises the lid to receive another charge of pollen.

46. Before proceeding further to consider how particular flowers are arranged to be helped by some particular sort or class of insects, and each in some peculiar way, we should contemplate the remarkable conclusion to which we are · brought. It seems to be this : — these flowers are so constructed that the pollen, however near the stigma, is somehow prevented from reaching it of itself, and then honey and other allurements are provided to tempt insects to come and convey the pollen to the stigma. And the various contrivances for hindering the pollen from reaching the stigma directly are excelled only by those for having it done in a roundabout way. So Nature appears to place obstacles in the way, and then to overcome the difficulty of her own making by calling in the aid of insects ! This is blocking the wheels with one hand and lifting the vehicle over the obstruction with the other. Or it is as if the wagoner of the fable, who prays Hercules to help him out of the mire, had bogged his team merely for the sake of calling upon Hercules. This is simply incredible. The explanation of one puzzle has brought in its train a greater puzzle still.

47. The solution of this puzzle is simple enough when once hit upon, although it has taken a long time to find it out. It not only makes everything plain as respects all these flowers, but also, as a true discovery should, clears up and explains a great many things besides. The explanation is, that

48. **Cross-Fertilization is aimed at.** The pollen was not intended to fertilize that same flower, but to be conveyed to some other flower of the same species. So insects, which had seemed to be needful only when the stamens and pistils are in separate flowers, or on separate plants, are quite as needful, — indeed, are more needful — where these organs stand side by side in the same blossom. The reason why crossing is advantageous, and in the long run necessary, is that

49. **Breeding-in-and-in is injurious.** Close-fertilization, that is, the fertilization of the seeds by pollen from the same flower, is very close breeding indeed. It is the next thing to no fertilization at all in plants, that is, to propagation by buds, — which may go on, as we know, for a long time : but it is not probable that any species could always continue in that way. Cultivators and stock-breeders are obliged to close-breed to keep a particular race of few individuals true and to heighten its desirable qualities. But sooner or later (in animals soon), more or less wide breeding is necessary to keep up vigor and fertility. Wide-breeding is

naturally secured by the structure itself in plants with separated flowers, — most completely in those which, like Willows, bear stamens and pistils upon different trees. And in the majority of plants which have perfect flowers it is commonly no less secured by arrangements of various kinds for excluding the pollen from its own stigmas, and having it conveyed to those of some other flower of the same species.

50. Comprehending now the full meaning of these curious arrangements, we may turn back to some of the flowers already noticed, to observe how exquisitely they are adapted to the purpose in view, and then advance to new and more varied illustrations.

51. **Cross-Fertilization in Iris** (Fig. 12). A little nectar is produced in the bottom of the tube or narrow cup of the blossom. The only access to it is a narrow channel leading down the united bases of the six divisions or leaves of the flower. Now the three inner of these are upright, with their tips curved inwards, shutting off all access in that quarter. But the three outer and larger divisions recurve and afford a convenient landing-place directly before the stamen and the overarching stigma. Here the bee alights. To reach and suck out the nectar with his proboscis will bring the head at least as low as the base of the anther. On raising his head to depart he sweeps with it the whole length of the anther and dusts it with the pollen now shedding. A little higher the shelf of stigma is hit, but only the outer face of it, which is smooth and does not take the pollen at all. Flying to the next blossom, the first thing which the pollen-powdered head of the bee strikes is the stigma, but this time on the upper face of the shelf or real surface of stigma, which takes some of the pollen brought into contact with it, and so is fertilized. Sinking lower, the head next brushes the anther downwards in entering for the nectar, then upwards in departing, and receives a fresh charge of pollen to be deposited upon the shelf of stigma of the next blossom visited, and so on.

52. **In Arethusa** (40, 45, Figs. 13 – 15). We have never seen bees or other insects about this flower; but it is plain from its structure that it cannot set seed without their help. As already described, the bee, or other insect of considerable size, can enter the blossom only in front; and the large and crested recurving petal offers a convenient landing-place. At the bottom of the narrowed cup of the flower a little nectar is produced, down to which the insect must reach its proboscis. In rising to escape, its head must strike the lower face of the over-

hanging shelf, which is stigma, and so sticky that any pollen it may chance to have brought would be left adhering there. As the head slips by, it must next hit the front edge or visor of the helmet-shaped anther, raise it on its hinge, and so allow one or more of the four loose pellets of pollen to drop out, or be brushed out by the insect's head, to which some of the pollen would stick. When the next flower is entered nothing is accomplished; but on departing, as before, any pollen on its head would be applied to the sticky shelf of stigma overhead, the lid then uplifted, and a fresh charge of pollen taken from this flower to be given to the next, and so on in succession.

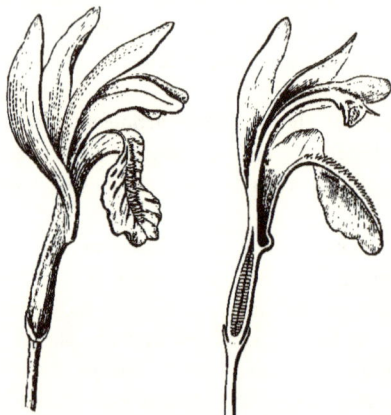

Fig. 13. Flower of Arethusa, entire. Fig. 14. A section lengthwise.

53. It is not unlikely that the pellets of pollen, as they fall out of the uplifted anther of Arethusa, may sometimes miss the insect's head, or fail to adhere to it, and so be lost. But this plan, or something like it, serves the purpose in the portion of the Orchis Family of which Arethusa is the representative. In others of that family the result is made surer by considerably different, more economical, and wonderfully curious arrangements, — such especially as those

Fig. 15. Diagram of the anther and stigma of Arethusa, put in upright position.

54. In **Orchises** and other plants of that particular tribe of the Orchis Family.   There is only one true Orchis in this country, and that not common, except northward.   And its arrangement for fertilization is not quite so readily understood as in those Orchises which are named by botanists *Habenaria*, of which we have many species.   Some of these are plentiful, such as the Fringe Orchises, either the purple, white, or yellow species. The Greater Green Orchis is not so common, but is taken for the present illustration on account of the size of its blossoms.   A reduced figure of it, with its two large round leaves spreading on the ground, and its spike of flowers rising between them on a naked stalk, is in the foreground of the vignette title, and a single blossom, of only twice the size of life, is represented in Fig. 16.

55. The peculiarities are mainly these : First, the better to attract certain insects and repay them for their service, a separate organ for the nectar — in this instance a long pouch or honey-tube — is attached to the flower. Then, to economize the pollen, the whole of it in each cell of the anther is done up in little packets or coarser grains, which are tied, as it were, to each other by delicate elastic threads, and all made fast by similar threads to the upper end of a central stalk. Finally, to make sure of its being taken by the insect and not dropped or lost in the carrying, the other end of this stalk bears a flat disk, commonly button-shaped, the exposed face of which is very sticky ; and this is placed just where it will be pretty sure to be attached to the head or proboscis of an insect that comes to drain the honey-tube. So that the insect, on rising from his meal, will probably carry off bodily the whole pollen of that flower (or of one of its anther-cells), and bestow it, or some of it, upon the next flower or flowers visited.

56. In this particular species, the front petal is, as usual, the insect's landing-place. The other petals are more arching than the front view of the flower in Fig. 16 represents, and obstruct access on all other sides. The long and narrow front petal turns downwards and allows convenient approach. Underneath hangs

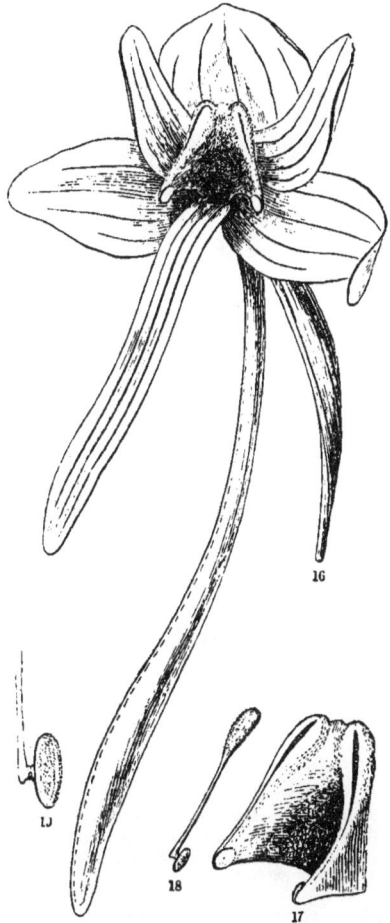

Fig. 16. Flower of Greater Green Orchis (Habenaria orbiculata). 17. Its stamen and stigma more enlarged. 18. One of the pollen-masses with its stalk and disk, equally enlarged. 19. Its disk and a part of the stalk more magnified.

the honey-tube, its mouth opening just behind the base of this petal.   Only the lower half of the tube, more enlarged and capacious, gets filled with nectar.   To

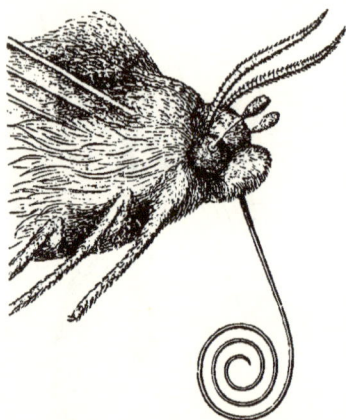

Fig. 20.   Side view of head of a moth (Sphynx drupiferarum), which has just extracted a pair of Orchis pollen-masses.

Fig. 21.   Front view of the same, with the pollen-masses in the position they soon take.   Both figures magnified to the same degree as is the Orchis flower in Fig. 16.

drain a cup which is about an inch and a half deep requires a long proboscis, much longer than any bee or wasp possesses.   Butterflies and moths are our only insects capable of doing it ; and one could predict from a view of the flower that the work is done by them.   In fact we have hardly a butterfly with proboscis long enough to reach the bottom of the cup : so we conclude that one of the Sphynxes or Night-moths, such as flock around the blossoms of the largest Evening-Primroses at dusk, is the proper helpmate of the Greater Green Orchis.   The Smaller Green Orchis is much like the Larger, but with honey-tube hardly an inch long.   This may be drained by many of our butterflies.   Some of these have been caught with a remarkable body attached to their great eyes, one on each eye ; on examination this body proved to be quite like that represented in Fig. 18, only smaller.   This body, as we have seen, is the pollen of one of the cells of an Orchis anther, with its stalk and sticky disk, the latter adhering to the insect's eye.   How did it get there ?

57. The centre of the flower (as in Fig. 16) is occupied by the one large anther, and by the concave stigma.   The anther is of two cells, which taper towards the front of the flower and diverge, in this species widely, and the whole space between the two diverging horns on the sides and the orifice of the honey-tube below is stigma, a broad patch of glutinous surface.   At the tip of each horn of the anther, facing forwards and partly inwards is the button-shaped, sticky disk.   Bring the point of a blunt pencil, or the tip of the little finger, or anything of the proper size,

down into the flower so as to press gently upon these disks for a moment; then withdraw it: the disks will stick fast, and the stalks with the pollen-mass be drawn out of the anther. Now the tip of the finger or the pencil is just in the position which the head of the large butterfly or moth would occupy when its proboscis is thrust deep into the honey-tube. In draining the nectar from the tube the insect's head is brought down close to its orifice, its large projecting eye on one side or the other, or on both at once, is pressed against the sticky button; and when the moth raises its head and departs, it carries away bodily one or both of the pollen-masses. With these the next flowers visited may be fertilized.

58. Except by the insect's aid as a carrier, secured by this most elaborate and wonderful contrivance, these Orchis flowers could never be fertilized. Close as the pollen is to the stigma, it evidently cannot reach it by any ordinary chance. And it would appear as if the obstacles were not effectually overcome even when a moth or butterfly is so ingeniously employed to convey the pollen from one blossom to another, which is plainly what is intended. For the position of parts is such that when the pollen-masses are extracted by the moth's head, they will stand pointing upwards and forwards, as shown in Fig. 20. The stalk is too stiff to allow them to subside by their own weight. So when the moth alights upon the next flower and thrusts its proboscis down its honey-tube, the pollen-masses it has brought would hit the anther, quite above the stigma, and effect nothing. But all this is accurately provided for. As may be seen by watching the pollen-masses when taken upon the point of a pencil, within from ten to thirty seconds their stalk turns downward, as if upon a joint between it and the adhering disk, bringing them into a position like that represented by a front view in Fig. 21. Now the pollen-masses will accurately strike the stigma!

59. In some Orchises, and where this adjustment is needful, the pollen-masses on the insect's head not only turn downwards but converge inwards, always in the way and to the degree necessary for their striking the stigma. In the larger Green Orchises, from which the illustrations are drawn, the sticky disk is almost parallel with the stalk of the pollen-mass at its lower end, and attached to it by a short intermediate joint, as shown in Fig. 18, and more magnified in Fig. 19. It is nearly the same in the Yellow and the White Fringed Orchises, which flower later in the season. In all these the disks face partly inwards, at considerable distance apart, and are stuck to the eye of the butterfly that visits them. In

others the disks are borne directly upon the end of the stalk, are generally closer together, and get applied to the front of the head, or sometimes to the proboscis of the insect.

60. When a pollen-mass, thus carried on the head of an insect, is brought into contact with the stigma, some of the pollen will cleave to its glutinous surface and be left there, the little threads that bind it to the stalk giving way; another portion will be left upon the stigma of the next flower visited, perhaps on the next also, and so nearly all the pollen be turned to good account. Sometimes the adhesion of the disk to the insect's eye is less strong than the threads that bind the grains to the stalk on the one hand, and than the adhesion to the stigma on the other. Then the whole pollen-mass is left upon the stigma of that flower, and its pollen taken in turn, to be exchanged for that of the next flower; and so on. In any case each blossom will be fertilized by the pollen of some other blossom, which is the end in view; and a more ingenious contrivance for the purpose cannot be imagined.

61. The student should see all these curious things with his or her own eyes, in order fully to comprehend and enjoy them. Once understood in our common wild Orchises, it will be equally interesting to find out how it is done, in more or less different and varied ways

62. **In other Orchids,** — whether wild ones, such as Ladies' Tresses, *Calopogon*, etc., or in those various and more gorgeous ones, mostly air-plants of tropical regions, which adorn rich conservatories. Some of these curiously resemble butterflies themselves, — either a swarm of them, as some of the smaller ones in a cluster on a long light stalk, fluttering with every breath of air; some are like a large, single, gorgeous, orange and spotted butterfly: *Oncidium Papilio*, for example (Fig. 22), which takes its name from the singular likeness, *Papilio* being Latin for butterfly; and *Phalænopsis*, a plant of which, greatly reduced in size, is represented on the vignette title-page (upper right-hand corner), with large white flowers, takes its name from its resemblance to a moth. Can the likeness be a sort of decoy to allure the very kinds of insect that are wanted for fertilizing these same flowers? Sometimes the strange shapes are not like insects; the flowers of *Stanhopea tigrina*, for example (figured at the top of the vignette title-page), resembling in color and form rather the head of a cuttle-fish than any known insect.

63. **In Lady's-Slipper, or Cypripedium,** the plan for securing fertilization is so dif-

ferent from that of any other of the Orchis Family as to need a separate description, but a very brief one must serve, as we have no figure ready. We refer to our

wild species; and first to the
yellow ones and to the large
white and pink one, *Cypripedium spectabile*, the Showy
Lady's-Slipper. Unlike other
Orchids, there are two stamens: the pollen is powdery, or
between powdery and pulpy,
and not very different from
that of ordinary flowers. As
it lies on the open anther in a
broad patch, it somehow gets
a film like a thin coat of sticky
varnish. The stigma is large,
flat, and somewhat trowel-
shaped, the face turned forwards and downwards: it is
supported on a stout style, to
which the anthers have grown
fast, one on each side. This
apparatus is placed just within
the upper part of the sac or

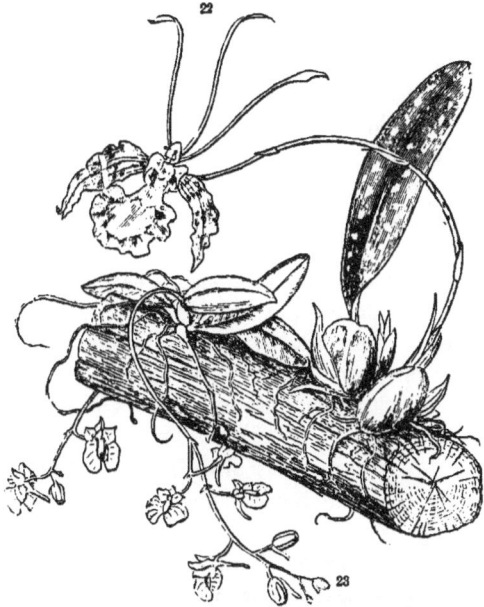

Fig. 22. Oncidium Papilio. Fig. 23. Comparettia rosea. Both
are Epiphytes, or Air-plants, and reduced in size.

slipper (rather like a moccason or buskin than a slipper), which gives name to the
flower. There are three openings into the slipper; a large round one in front,
and the edges of this are turned in, after the fashion of one sort of mouse-trap;
two small ones far back, one on either side, directly under each anther. Flies and
the like enter by the large front opening, and find a little nectar apparently be-
dewing the long hairs that grow from the bottom of the slipper, especially well
back under the overhanging stigma. The mouse-trap arrangement renders it difficult for the fly to get out by the way it came in. As it pushes on under the
stigma it sees light on either side beyond, and in escaping by one or the other of
these small openings it cannot fail to get a dab of pollen upon its head, as it
brushes against the film with which the surface is varnished. Flying to the next

blossom and entering as before, as the insect makes its way onward, it can hardly fail to rub the pollen-covered top of its head against the large stigma which forms the roof of the passage.   The stigma of every other Orchid is smooth and glutinous.   This is merely moist and finely roughened : the roughness comes from very minute projections, all pointing forwards, so that the surface may be likened to that of a wool-card or of a rasp on a very fine scale.   So, as the insect passes under, the film of pollen is carded or rasped off its head by the stigma and left upon it ; and when the fly passes out it takes a fresh load of pollen on its head with which to fertilize the next flower.   This mode of action we first predicted from an inspection of the flower and a simple experiment.   It has since been confirmed by repeated observations.   The early-flowering and purple Stemless Lady's-Slipper differs from the others in having its larger slipper or sac pendent, and with a long slit in front, instead of a round open orifice ; the two lips of the slit are mostly in contact, but the fly may readily push its way in ; the way of exit is more open than in the other species.

64. **In Asclepias or Milkweed.**   Now and then the rough legs of butterflies and bees are found to be encumbered with bodies sticking to them which resemble the

pollen-masses of Orchids ; but there is always a pair of them, of waxy appearance, hanging by a curved stalk from a dark-colored disk, if it may be so called, which is not button-shaped.   These are the pollen-masses of Milkweed, carried off by insects alighting on the flower to suck the nectar from five little cups, and, sticking fast to their legs or feet, are so carried from flower to flower.   Fig. 24 shows a pair of

Fig. 24. Pair of pollen-masses of Milkweed.

them.   Milkweeds are like Orchids in this respect only.   Their flowers are very different and peculiar, not readily to be explained except with the plant itself in hand ; but insects are equally necessary to fertilize them.

65. How ordinary blossoms are cross-fertilized by insects passing continually from flower to flower will be obvious enough after these explanations.   But observing eyes will detect many curious arrangements in the commonest plants, now that the way is pointed out.   A few may be described.

66. **In Barberry-blossoms** there is a remarkable peculiarity.   We have learned, in the first chapter, that certain plants are endowed with the power of moving some part freely in order that they may climb.   Barberry-blossoms have a movement upon irritation, which has long been familiar as a mere curiosity, but which we

now begin to understand the meaning of. It is turned to account in fertilization. The six stamens surround a pistil, but diverge away from it, as if to be sheltered, one under each of the concave or arching petals. There they remain unless touched, as with a pin or any other body, at the base of the filament on the inside; then the stamen starts forward suddenly, as with a jerk, into an erect position. Not far enough forward, however, for the anthers to hit the stigma; indeed, the filament is not quite long enough for that. Now the anther opens in an unusual way, namely, by trap-doors, one on each side (as shown in Fig. 25), letting the pollen drop out. Barberry-blossoms are visited by honey-bees and by smaller flying insects; in the common Barberry the flowers are hanging. A touch by the proboscis of a bee hovering underneath causes the stamens in turn to spring forward suddenly, and to shower the insect plentifully with their pollen. Some of this may be applied immediately to the button-shaped stigma of that very flower; but some would surely be carried to the stigma of the next flowers visited, and so on. In species with upright flowers, the pollen will dust the proboscis and head of the bee, or of smaller insect crawling to the bottom for the nectar there; and in entering a subsequent blossom it must needs brush this pollen against its stigma.

Fig. 25. Stamen of Barberry: anther opening by trap-doors.

67. In Kalmia (*American Laurel*, and equally in the smaller species, namely, Sheep Laurel or Lambkill, and in the earlier-flowering Glaucous Laurel of the bogs), a mechanical instead of a vital movement is turned to similar account. The singular structure of the blossom has long been known; the operation of it is only now understood.

68. This is the plan of it. Ten stamens with slender filaments surround a still longer style : the tip of the style is the stigma, which the pollen is somehow to reach. But the anthers in the flower-bud lie in as many pouches in the sides of the corolla (Fig. 26). When the corolla opens and takes its saucer-shaped form, the anthers remain lodged in the pouches, so the filaments are bowed back and become so many springs (Figs. 27, 28). If untouched the springs generally remain set until the corolla begins to fade : by that time the filaments lose their elasticity and become flabby also. If we jostle them, however, by a somewhat rude touch when the flower is in fresh condition, so as to liberate the anther, the filaments straighten elastically and suddenly, and generally curve over in the opposite direction. As they fly back they discharge a quantity of pollen.

3

Take notice that these anthers do not open by trap-doors, like those of Barberry, nor by long slits as in most flowers. As in most of the Heath Family (to which Kalmia belongs), they consist of a pair of sacs, side by side, which open by a round hole at the top (see Fig. 29). So, when the bowed filament is set free and flies forward, the grains of pollen in the anther are projected, like shot from a child's pea-shooter. A bit of whalebone, to the end of which two pieces of quill filled with small shot are made fast, is not a bad representation of one of these stamens. This really must be a contrivance for discharging pollen at some object. If the stigma around which the stamens are marshalled be that object, the target is a small one, yet some one or more of the ten shots might hit the mark. But the discharges can hardly ever take place at all without the aid of an insect. Bees are the insects thus far observed to frequent these flowers; and it is interesting to watch the operations of a bumble-bee upon them. The bee, remaining on the wing, circles for a moment over each flower, thrusting his proboscis all round the ovary at the bottom; in doing this it jostles and lets off the springs, and receives upon the under side of its body and its legs successive charges of pollen. Flying to another blossom, it brings its pollen-dusted body against the stigma, and, commonly revolving on it as if on a pivot while it sucks the nectar in the bottom of the flower-cup, liberates the ten bowed stamens, and receives fresh charges of pollen from that flower while fertilizing it with the pollen of the preceding one. This account is founded on the observations of Professor Beal, of Michigan, who also states that when a cluster of blossoms is covered with fine gauze, no stamen gets liberated of itself while fit for action, and no seed sets.

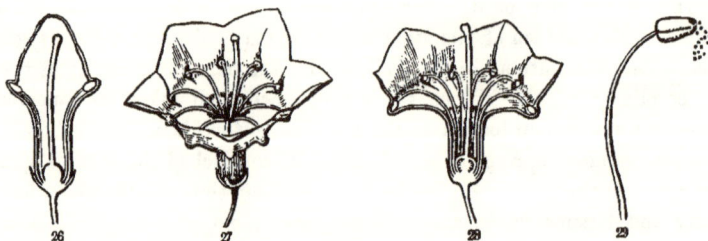

Figs. 26 - 29. Flower of American Laurel, Kalmia latifolia. 26. Flower-bud divided lengthwise. 27. Open flower. 28. Section of same, lengthwise. 29. A stamen enlarged, discharging pollen from the two holes at the top.

69. One might doubt whether such movements as those of the stamens of Barberry and of Kalmia were really intended for the use here assigned to them. But they serve this purpose, unquestionably, and we can think of no other. Now there is a flower of a tropical Orchid, cultivated in some conservatories (named *Catasetum*), in which a movement under irritation (analogous to that of the Barberry-stamen) and one of elasticity (like that of Kalmia) are combined in one apparatus, — one so elaborate and special that nobody can doubt that it is a contrivance for this particular purpose. It cannot well be described here without numerous figures and much detail. But the amount of it is, that a sensitiveness of two slender and partly crossed arms, which the moth or other large insect must hit in reaching the flower-cup, liberates a pollen-mass which is set as a spring, and lets it fly like a catapult; it hits the head of the insect at some distance, disk-end foremost, and sticks fast to it, in proper position to be applied to the stigma of the next proper flower visited.

70. Returning to flowers of ordinary structure, and of familiar kinds, two particular arrangements for insuring cross-fertilization in perfect flowers must be briefly noticed. The commonest is that of

71. **Dichogamous Flowers.** *Dichogamy* is the name given to the case in which the stamens and the stigmas of the same blossom come to perfection at different periods. That is, the anthers mature and discharge their pollen in some plants before the stigma is ready to receive it, in others only after the stigma has withered. Either way, the pollen that fertilizes and the stigma that is fertilized can never belong to the same blossom.

72. **In the Common Plantain** of our dooryards and waysides, *Plantago major*, and in the English Plantain, or Ripple Grass (*P. lanceolata*) of the fields, this is familiarly illustrated. The style projects from the apex of the closed bud, ready to receive pollen from other flowers a day or two before its stamens are hung out upon their slender filaments, to furnish pollen for other flowers, — not for their own, the stigma of which is by that time dried up. Plantain-flowers, however, produce no nectar, and are neither fragrant nor brightly colored; so they are not visited by insects, but are left to the chance of the conveyance of the pollen by the wind. It is the same with many Grasses and Grains, only in reverse order. Their anthers hang out on their slender filaments one morning, and the feathery stigmas of that blossom not until the next morning; and the wind is the pollen-carrier.

73. In Figwort or Scrophularia, and in many other flowers of which this may serve as an example, the work is done with much saving of pollen by calling in

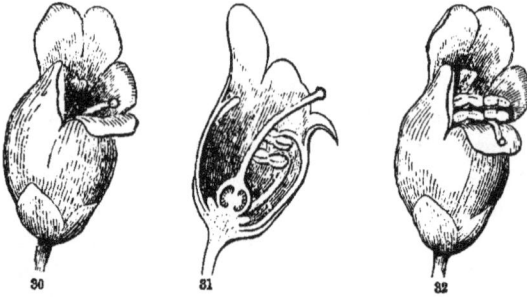

the aid of insects. Fig. 30 is an enlarged representation of one of the flowers, as it appears throughout the day of opening. The style projects from the gorge of the corolla, presenting the stigma just over the front edge. The stamens are out of sight and reach, and not yet ready: they lie recurved below, as shown in Fig.

Fig. 30. Flower of Scrophularia nodosa, the first day. 31. Inside view of it, the front half cut away. 32. Flower as it appears on the second day.

31. A day or two later the flower appears as in Fig. 32: the style is flabby or withering, and the stigma dried up; the stamens have straightened their filaments, and have brought up the four now opened anthers above the front edge of the corolla, where the stigma was the day before. The bottom of the corolla-cup contains some nectar. Honey-bees are attracted by it. When they visit a flower in the state of Fig. 32, alighting as they do on the front lip, they get the chest and legs well dusted with pollen, none of which has acted upon its own stigma; for that was dry and effete before these anthers opened. When the bee passes to a freshly expanded flower, such as Fig. 30, the parts covered with pollen are sure to be brought against the fresh and active stigma, which cannot have possibly been touched by any pollen of that flower, its anthers being still immature and hidden below.

74. In some other Flowers the pollen is conveyed from an earlier to the stigma of a later blossom, the anthers maturing and shedding their pollen before the stigma is ready to receive any. A beautiful case of the sort, in which a movement comes conspicuously into play, may be seen in *Clerodendron Thompsoniæ*, a climbing shrub from tropical Africa which blooms in our conservatories. Four stamens with very long filaments and an equally long and slender style are rolled up together in the corolla-bud. When this expands, the stamens straighten out nearly in the line of the tube of the corolla, and their anthers open: the style has bent so far forwards as to point downwards, and the stigma is not yet ready for pollen, its two branches being united. So a butterfly, in the act of drawing nectar from this flower, will get the under side of its body dusted with pollen, but will not come near the reflexed and still immature style. But in a flower a day older, the stamens are found to be coiled up (the opposite way from what they were in the bud) and turned down out of the way, bringing the anthers nearly where the stigma was the day before; while the style has come up to where the stamens were the day before, and its stigma with branches outspread is now ready for pollen, — is just in position and condition for being dusted with the pollen which the butterfly has received from the anthers of an earlier blossom.

75. Campanulas and Sabbatias also mature their anthers and shed their pollen long before the stigmas open so as to receive any; they, too, are fertilized by insects carrying pollen from an earlier to a later flower. To understand how it is done in each particular case the flowers themselves should be studied in the field and garden.

76. **Dimorphous Flowers,** that is, flowers of two kinds as to length or position of stamens and pistil, but both sorts perfect, remain to be considered. In these the difference is only in the stamens and pistil, usually merely in their relative length, and very likely to be noticed only by the attentive observer. A good case of this may readily be seen

77. **In Houstonia.** The commonest species, the little blue-eyed *Houstonia cœrulea*, looks up to us from every low meadow in spring as soon as the turf gets dry enough to set foot upon. In different patches of it, some flowers will show the tips of the four stamens slightly projecting; as many others will show the two stigmas only. The two kinds are always in different patches; all that come from the same seed being alike. The sort that shows the tips of the anthers (as in Fig. 33, and with

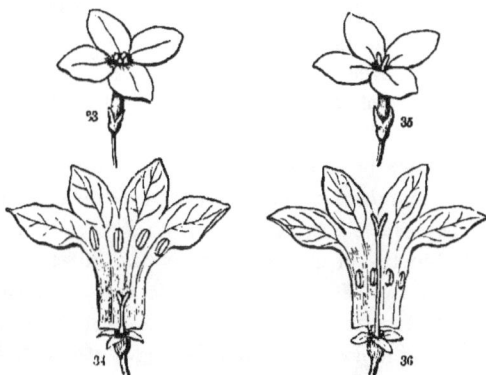

Figs. 33, 35. The two sorts of flower of Houstonia cœrulea. 34 and 36. Same more enlarged, with corolla divided and laid open.

corolla divided and spread open in Fig. 34) has a short style, which brings the two stigmas up to near the middle of the tube of the corolla. The sort that shows the stigmas projecting (as in Figs. 35 and 36) has the style long enough to bring them up just to the place which the anthers occupy in the other flower; but its anthers are placed as low down in the tube as the stigmas are in the first flower. The little Partridge Berry of the woods has its flowers of two sorts, on the same plan: and among garden flowers it may be seen in Primroses. But it is to be noted that this plan occurs only in flowers that are frequented by insects.

78. In the *Houstonia*, small insects, feeding by a proboscis, passing from flower to flower, take from the high-stamened one (Figs. 33, 34) some pollen upon the face, as it is brought down close to the orifice of the corolla when the proboscis is thrust to the bottom for the nectar there. When the insect passes to another flower of the same sort, it merely gets its face smeared with a little more pollen. But when it visits a long-styled flower (such as Figs. 35, 36) and brings its head down to the orifice, it will apply some of this pollen to the stigmas, which are exactly in the position to receive it. So the high anthers are to fertilize the high stigmas. How about the low stamens and low stigmas, when the insect flies from a flower of the second sort to one of the first, as it is quite as likely to do? Why, the insect's proboscis, as it explores that flower, gets dusted with the pollen of the low anthers, and this pollen is neatly carried and applied to the similarly placed stigma of the other kind of flower. So much for dimorphous flowers. There are even

79. **Trimorphous Flowers,** that is, perfect flowers of three sorts arranged to co-operate in this way. One case at least was discovered by the most sagacious investigator of this whole class of subjects (Mr. Darwin), in a kind of Loosestrife (*Lythrum Salicaria*), and there is something nearly like it in another bog plant of the Loosestrife Family, *Nesœa verticillata*. There are three lengths of style and three lengths of stamens, two of the latter in each sort of flower, the stamens being in two sets. Bees suck the flowers of this Loosestrife. In doing so, the longest stamens rub their pollen against the lower and hinder part of the body and the hind legs; the middle-length stamens, between the front pair of legs; the shortest stamens, against the proboscis and chin. When they fly to other flowers, the very parts that are dusted with long-stamen pollen rub against the stigma of the long style; those dusted with that of middle-length stamens, against the stigma of middle-length style; those with that of short stamens, against

the stigma of the shortest style, — each to each. Not only is the pollen, through such wonderful arrangements, so distributed as to secure cross-fertilization, but the end is further secured by a

80. **Preference of Stigma for Pollen of other Flowers than its own.** In dimorphous and trimorphous flowers, such as have just been described, it has been ascertained that if pollen is placed upon the stigma of the same blossom, or even on that of another blossom of the same sort, it takes little or no effect. There are cases where the stigma gets naturally covered with its own-flower pollen without setting seed, but when touched with the pollen of another flower it seeds perfectly. This explains, at length, the remarkable thing (described in paragraph 37) that the blossoms of Peas, Beans, and of Dicentra or Bleeding-heart and the like, generally set little or no seed when insects are excluded, although the parts are so disposed that the stigma must be dusted by the pollen of the stamens enclosed with it. Why even such flowers need the aid of insects is now clear. This preference of pollen for other than its own blossom, however, is strictly

81. **Within the limits of the Species.** The pollen which is conveyed to the stigma of a different species is inactive and without result, in all but species that are pretty nearly related, and in many of these. Apple-blossom pollen, for instance, does not fertilize pear-blossoms, and vice versa. Cross-breeding among flowers of the same species is the rule, — among different species the exception. It may be done, however, to a certain extent, but always with more difficulty; it rarely occurs in nature left to itself. Crossing of species produces *Hybrids :* by recourse to it gardeners and florists greatly diversify certain flowers and fruits ; for the new sorts produced inherit from both parents : the cultivator aims at originating and preserving those that combine the most desirable qualities of both parents.

82. **Advantage of Perfect Flowers.** The greater number of species, and far the greater number of those that are visited by insects, are perfect, that is, with stamens and pistil in the same blossom. Yet separated flowers would seem best for the end in view, cross-fertilization in them taking place of necessity. But, with insects to assist, it is better, that is, more economical, to have perfect flowers ; for, while the crossing is equally secured, both flowers produce seed. " The economy of Nature " of which we read is something more than a figure of speech.

83. The reciprocity of flower and flower, and of insects and flowers, is something admirable. Insects pay liberal wages for the food which flowers provide for them. The familiar rhymes of Dr. Watts directed the attention of young people

to the bee visiting the flower as a model of industry. With a slight change of a couplet, adapting it to our present knowledge and to the lesson of mutual helpfulness, we may read : —

> How doth the little busy bee
> Improve each shining hour,
> While gathering honey day by day,
> To fertilize each flower.

84. Such are the principal modes, thus far known (and when these are understood watchful eyes may discover other equally curious cases), in which flowers are prevented from breeding in and in, either wholly or to such extent as to keep up the vigor of the species. Such are some of the ways in which flowers are adapted to insects, and no doubt insects to flowers, for this end. Plants, destitute of the locomotion and volition which animals, at least the higher animals, enjoy, have the lack made up to them in these subtle and very various contrivances, by which the volition and locomotion of insects are made to serve them, even to secure their very existence. For, to say that these plants could continue to flourish without such aid is tantamount to saying that these multifarious, elaborate, and exquisite arrangements are superfluous, — which is past all belief.

85. It is equally past belief that they are undesigned or accidental. No one has been able to describe them except in language which assumes that they are *contrivances, adaptations for particular purposes,* and the like ; and where many of them are best described they are said to " transcend in an incomparable degree the contrivances and adaptations which the most fertile imagination of the most imaginative man could suggest, with unlimited time at his disposal." Now, no matter whether or not the flowers themselves with all these structures have been perfected step by step, through no matter how long a series of natural stages, — if these structures and their operations, which so strike the mind of the philosopher no less than of the common observer that he cannot avoid calling them contrivances, do not argue intention, what stronger evidence of intention in Nature can there anywhere possibly be ? If they do, such evidences are countless, and almost every blossom brings distinct testimony to the existence and providence of a Designer and Ordainer, without whom, we may well believe, not merely a sparrow, not even a grain of pollen, may fall.

CHAPTER III.

HOW CERTAIN PLANTS CAPTURE INSECTS.

86. THIS is not a common habit of plants. Insects are fed and allowed to depart unharmed. When captures are made they must sometimes be purely accidental and meaningless; as in those species of *Silene* called Catch-fly, because small flies and other weak insects, sticking fast to a clammy exudation of the calyxes in some species, of a part of the stem in others, are unable to extricate themselves and so perish. But in certain cases insects are caught in ways so remarkable that we cannot avoid regarding them as contrivances, as genuine *flytraps*.

87. **Flower-Flytraps** are certainly to be found in some plants of the Orchis Family. One instance is that of Cypripedium or Lady's-Slipper, which, being a contrivance for cross-fertilization, is described in the foregoing chapter (paragraph 62). Here the insect is entrapped for the purpose of securing its services; and the detention is only temporary. If it did not escape from one flower to enter into another, the whole purpose of the contrivance would be defeated. Not so, however, in

88. **Leaf-Flytraps.** These all take the insect's life, — whether with intent or not it may be difficult to make out. The commonest and the most ambiguous leaf-flytraps are

89. **Such as Pitchers,** of which those of our *Sarracenia* or Sidesaddle-flower are most familiar. Fig. 37 represents one leaf, and a section of another, of the species most common in our bogs, especially at the North; and the vignette title-page, at bottom on the right hand, shows the longer and more tubular pitchers of another species of the Southern States. *S. flava*, a common yellow-flowered species from Virginia southward, has them so very long and

Fig. 37. Leaf of the common Sarracenia purpurea, and one cut across.

narrow, that they are popularly named *Trumpets*. In these pitchers or tubes water is generally found, sometimes caught from rain, but in other cases evidently furnished by the plant, the pitcher being so constructed that water cannot rain in : this water abounds with drowned insects, commonly in all stages of decay.   One would suppose that insects which have crawled into the pitcher might as readily crawl out ; but they do not, and closer examination shows that escaping is not as easy as entering.   In most pitchers of this sort there are sharp and stiff hairs within, all pointing downward, which offer considerable obstruction to returning, but none to entering.

90. Why plants which are rooted in wet bogs or in moist ground need to catch water in pitchers, or to secrete it there, is a mystery, unless it is wanted to drown flies in.   And what they gain from a solution of dead flies is equally hard to guess, unless this acts as a liquid manure.

91. Into such pitchers as the common one represented in Fig. 37 rain may fall ; but not readily into such as those of the vignette title already referred to, —

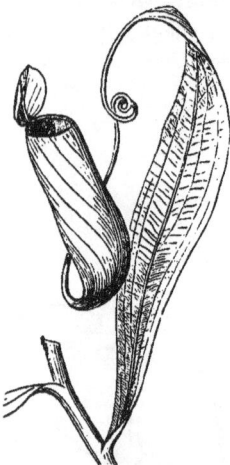

Fig. 38. Leaf-tendril and pitcher of Nepenthes.

not at all into those of the Parrot-headed species, *S. psittacina* of the Southern States, for the inflated lid or cover arches over the mouth of the pitcher completely.   This is even more strikingly so in *Darlingtonia*, the curious Californian Pitcher-plant lately made known and cultivated : in this the contracted entrance to the pitcher is concealed under the hood and looks downward instead of upward ; and even the small chance of any rain entering by aid of the wind is, as it were, guarded against by a curious appendage, resembling the forked tail of some fish, which hangs over the front.   Any water found in this pitcher must come from the plant itself.   So it also must in the combined

92. **Pitcher and Tendril of Nepenthes.**   These Pitcher-plants are woody climbers, natives of the Indian Archipelago, and not rarely cultivated in hot-houses, as a curiosity.   One is shown on the vignette title, right-hand side, and their way of climbing is mentioned in the foregoing chapter (19).   Some leaves lengthen the tip into the tendril only ; some of the lower bear a pitcher only ; but the best developed leaves have both, — the tendril for climbing, the

pitcher one can hardly say for what purpose. The pitcher is tightly closed by a neatly fitting lid when young; and in strong and healthy plants there is commonly a little water in it, which could not possibly have been introduced from without. After they are fully grown the lid opens by a hinge; then a little water might be supposed to rain in. In the humid sultry climates they inhabit it probably does so freely, and the leaves are found partly filled with dead flies, as in our wild Pitcher-plants.

93. The drowning of insects in plant-pitchers is of course an accidental occurrence, and any supposed advantage of this to the plant may be altogether fanciful. But we cannot deny that the supply of liquid manure may be useful. Before concluding that they are of no account, it may be well to contemplate other sorts of leaf-fly traps.

94. **Sundew as a Fly-catcher.** All species of Sundew (*Drosera*) have their leaves, and some their stalks also, beset with bristles tipped with a gland from which oozes a drop of clear but very glutinous liquid, making the plant appear as if studded with dew-drops. These remain, glistening in the sun, long after dew-drops would have been dissipated. Small flies, gnats, and such-like insects, seemingly enticed by the glittering drops, stick fast upon them and perish by starvation, one would suppose without any benefit whatever to the plant. But in the broad-leaved wild species of our bogs, such as the common Round-leaved Sundew (figured, much reduced in size, at the foot of the vignette title, toward the right), the upper face and edges of the blade of the leaf bear stronger bristles, tipped with a larger glutinous drop, and the whole forms what we must allow to be a veritable fly-trap.

95. For, when a small fly alights on the upper face, and is held by some of the glutinous drops long enough for the leaf to act, the surrounding bristles slowly bend inwards so as to bring their glutinous tips also against the body of the insect, adding, one by one, to the bonds, and rendering captivity and death certain. This movement of the bristles must be of the same nature as that by which tendrils and some leafstalks bend or coil. It is much too slow to be visible except in the result, which takes a day or two to be completed. Here, then, is a contrivance for catching flies, a most elaborate one, in action slow but sure. And the different species of Sundew offer all gradations between those with merely scattered and motionless dewy-tipped bristles, to which flies may chance to stick, and this more complex arrangement, which we cannot avoid regarding as intended for

fly-catching. Moreover, in one of our species with longer leaves (*D. longifolia*) the blade of the leaf itself incurves (as an intelligent lady has observed), so as to fold round its victim !   ·

96. Another and a most practised observer, whose observations are not yet published, declares that the leaves of the common Round-leaved Sundew act differently when different objects are placed upon them. For instance, if a particle of raw meat be substituted for the living fly, the bristles will close upon it in the same manner; but to a particle of chalk or wood they remain nearly indifferent. If any doubt should still remain whether the fly-catching in Sundews is accidental or intentional, — in other words, whether the leaf is so constructed and arranged in order that it may capture flies, — the doubt may perhaps disappear upon the contemplation of another and even more extraordinary plant of the same family with the Sundew, namely,

97. **Venus's Flytrap, or Dionœa muscipula.** This plant abounds in the low savannas around Wilmington, North Carolina, and is native nowhere else. It is not very difficult to cultivate, at least for a time, and it is kept in many choice conservatories as a vegetable wonder.

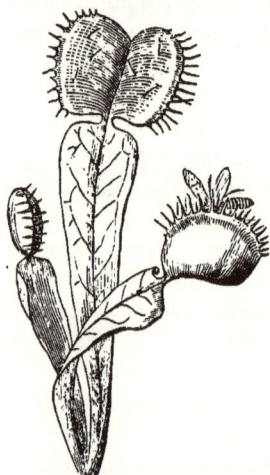

98. The trap is the end of the leaf (see Figs. 39, 40). It is somewhat like the leaf of Sundew, only larger, about an inch in diameter, with bristles still stouter, but only round the margin, like a fringe, and no clammy liquid or gland at their tips. The leaf folds on itself as if hinged at the midrib. Three more delicate bristles are seen on the face upon close inspection. When these are touched by the finger or the point of a pencil, the open trap shuts with a quick motion, and after a considerable interval it reopens. When a fly or other insect alights on the surface and brushes against these sensitive bristles, the trap closes promptly, generally imprisoning the intruder. It closes at first with the sides convex and the bristles crossing each other like the fingers of interlocked hands or the teeth of a steel-trap, as in the side figures of Fig. 39. But soon the sides of the trap flatten down and press firmly upon the victim; and it now requires a very

Fig. 39. Leaves of Dionæa or Venus's Flytrap, the trap of the larger one wide open.

considerable force to open the trap. If nothing is caught the trap presently reopens of itself and is ready for another attempt. When a fly or any similar insect is captured it is retained until it perishes, — is killed, indeed, and consumed; after which it opens for another capture. But after the first or second it acts sluggishly and feebly, it ages and hardens, at length loses its sensibility, and slowly decays.

99. It cannot be supposed that plants, like boys, catch flies for pastime or in objectless wantonness. Living beings though they are, yet they are not of a sufficiently high order for that. It is equally incredible that such an exquisite apparatus as this should be purposeless. And in the present case the evidence of the purpose and of the meaning of the strange action is wellnigh complete. The face of this living trap is thickly sprinkled with glands immersed in its texture, of elaborate structure under the microscope, but large enough to be clearly discerned with a hand lens; these glands, soon after an insect is closed upon, give out a saliva-like liquid, which moistens the insect, and in a short time (within a week or two) dissolves all its soft parts, — digests them, we must believe; and the liquid, with the animal matter it has dissolved, is re-absorbed into the leaf! We are forced to conclude that, in addition to the ordinary faculties and function of a vegetable, this plant is really carnivorous.*

100. That, while all plants are food for animals, some few should, in turn and to some extent, feed upon them, will appear more credible when it is considered that whole tribes of plants of the lowest grade (Mould-Fungi and the like) habitually feed upon living plants and living animals, or upon their juices when dead. An account of them would make a volume of itself, and an interesting one. But all goes to show that the instances of extraordinary behavior which have been

---

* Ellis, who first described the *Dionæa* in full, and gave it this name noticed the liquid secretion and the glands that produce it, but thought that it was given out while the trap was open and as a lure to insects: he expressed his belief that the leaves caught insects for the purpose of nutrition. Linnæus appears to have doubted this; he omitted all account of the fluid, and gave a more humane, but incorrect, version of the plant's behavior, stating that the trap holds the insect only while it struggles, but releases it on becoming quiet: and this statement has been commonly adopted. Elliott merely copied the description by Linnæus. The Rev. Dr. M. A. Curtis of North Carolina (just deceased) gave a more correct account about thirty years ago. Recently Mr. William M. Canby of Delaware has published some very interesting observations and experiments; which show that the liquid is a sort of gastric juice, exuded after the capture. He also fed the leaves with morsels of raw beef, and found that these in most instances were mainly dissolved in the juice, which then disappeared, evidently by absorption. Similar observations and experiments made by Mr. Darwin are still unpublished.

recounted in these chapters are not mere prodigies, wholly out of the general order of Nature, but belong to the order of Nature, and indeed are hardly different in kind from, or really more wonderful than, the doings of many of the commonest plants, which, until our special attention is called to them, ordinarily pass unregarded.

Fig. 40. Venus's Flytrap: Dionæa.